CRITERIA FOR THE SELECTION OF TEMPORARY WASTE STORAGE FACILITIES

Published by CBRNE Ltd

Title:	Criteria for the selection of temporary waste storage facilities	
Date:	August 29, 2014	
Author(s):	N J Hale, Dominic Kelly	CBRNE Ltd

This project has received funding from the European Community's Seventh Framework Programme. The views expressed in this document are purely those of the writer and may not in any circumstances be regarded as stating an official position of the European Community.

Front Cover Design by: Carolyn Smith BA (Hons) Ind Des MFA - CBRNE Ltd Design Director

Contents

1. Executive Summary

During the Recovery Phase[1] after a Chemical, Biological or Radiological (CBR) type incident which contaminated assets in locations under either private or public ownership, temporary storage of waste could provide a "buffer" such that recovery can proceed unhindered by the logistics of final waste disposal.

The European Waste Framework Directive and related legislation provides a framework by which temporary storage may be permitted with reduced administrative burden compared to longer term or semi-permanent storage.

The purpose of this document is to present criteria that organisations may use to assess their own facilities[2] to identify if they are likely to meet acceptance and approval from the relevant authorities as temporary waste storage facilities[3]. As there is currently no formalised process for acceptance of temporary waste storage facilities, consideration of these criteria will assist in securing appropriate arrangements, but of course cannot guarantee acceptance. A typical user of the criteria that are presented could be a Facilities Manager, who is seeking to find a suitable storage location for wastes that are being produced by his remediation subcontractor, pending identification of a disposal route. Extreme incidents, or those involving contamination of open places and public assets, may require additional and more detailed considerations than those presented here, although the criteria that are presented will still provide a robust starting point.

An examination of a number of example scenarios shows that temporary storage could range from:

    i)   simple expedients like wrapping in polythene or placing in drums which are then placed within ISO freight containers in a secure area...to,

    ii)   packaging of wastes and transporting them to a disused hangar/ suitable building on a secure site.

A consideration of the timelines involved in a remediation project and the plans presented in PRACTICE Deliverable D5.12 "Remediation Plans and Templates" indicates that planning for temporary waste storage needs to take place as early in a remediation project as is practicable. In the planning process it may also be possible to identify potential temporary facilities in advance or to arrange 'call-off' contracts or framework agreements with suitable providers.

Criteria

Four Key Criteria are identified, namely that the storage site is safe, practicable, acceptable and cost effective. The criteria are not completely independent; assessment of a particular site against them may require an iterative review as various sub-options are considered.

By considering the issues associated with each of the Key Criteria and in particular the one of Safety, a number of sub-criteria have been identified. Previous studies on specific waste storage

---

[1] See PRACTICE Deliverable D3.1 "Survey Methodology"
[2] Or those made available to them
[3] The selection, licensing and justification of sites for permanent storage is not covered by this document, it is the subject of various EC Directives.

sites and best practice guidance from publish standards – such as those for storage of radioactive and biological wastes – have also been used to identify sub-criteria.

For the consideration of the criterion of safety, a risk based approach is recommended; i.e. the acceptability of the performance of a candidate site should be judged based upon the frequency and likelihood of an adverse situation arising. This approach is widely used across Europe.

The sub-criteria and associated guidance notes are presented in an Annex to the main report.

The importance of the inclusion of stakeholders in the assessment of candidate sites, against the criteria, has been identified.

2. Introduction

Following an incident involving the release of CBR materials which contaminate assets at a site, after the emergency services have departed, the site owner may be left with the task of remediating[4] the site and its assets. PRACTICE deliverable D5.12 "Remediation Plans and Templates" provides guidance to such organisations regarding the types of documentation and procedures that should be produced to ensure acceptance of these projects by stakeholders. One of the factors that may have to be considered is the temporary storage[5] of wastes produced during the remediation, pending their final disposal or destruction. Such temporary storage could be required, for example, to[6];

- Provide buffer capacity pending the identification of a final waste treatment and disposal site (which may have a long lead time – given the uncommon or unusual nature of some of the contaminants of concern here)

- Permit more rapid completion of the remediation work and the return to normality[7]

- Provide buffer capacity so that the site can be kept clear to allow cleaning and restoration activities to proceed as quickly as possible.

- Minimise the need to handle/transport waste repeatedly, thus maximising the economics of transportation and reducing the associated nuisance and disturbance this may cause.

- Provide facilities in which wastes can be progressively segregated, and pre-treated if appropriate.

- Provide a flow of waste material into the ultimate treatment and/or disposal facilities that can be controlled and adjusted to match the processing capacity of the disposal site(s).

- Monitor, track and record all the different types and amounts of waste that are recovered.

Where the temporary storage area is either on the site where the wastes are produced or is on another site owned by the producer there are provisions in EU legislation (see WFD - the Waste Framework Directive) that permit these with reduced administrative burden – although some form of permit may still be required. Where the sites do not fall into these definitions, more detailed and complex justifications may need to be prepared; the extent of which is a matter for the country's government agencies to determine. In either case, a consistent set of criteria must be addressed in order to demonstrate that the sites will be safe and the criteria presented here will be appropriate.

The purpose of this document is to present criteria that organisations may use to assess their own facilities (or those made available to them) to identify if they are likely to meet acceptance and

---

[4] Remediation encompasses all of the steps that an Organisation must undertake to return their assets to the state chosen by them. See PRACTICE Deliverable D5.12 (Remediation Plans and Templates) for a fuller discussion.

[5] Temporary storage is defined as storage by the producer – at sites under their control - pending final disposition to a waste disposal site. It is a term defined within European waste regulations such as WFD.

[6] See for example SLR Consulting, UNEP/OCHA and MoESDR

[7] For example if a large shopping complex was contaminated then it might be desirable and practicable to rapidly contain the contaminated assets (e.g. in polythene bags) and remove them from the site such that the centre can be remediated, restocked and re-opened with the minimum of delay and financial impact.

approval from the relevant authorities as temporary waste storage facilities. Consideration of these criteria does not itself guarantee acceptance, but it will assist with it.

A typical user could be a Facilities Manager, perhaps supported by a Health and Safety Manager, who is seeking to find a suitable storage location for wastes that are being produced by his remediation subcontractor, pending identification of a disposal route.

Section 3 of this document presents a timeline which shows the potential sources of waste during a remediation project. Section 4 provides some simple example incident scenarios to provide an indication of the potential waste issues and some potential temporary storage solutions.

The criteria themselves are presented in tabular form, along with explanatory comments, in Section 5.

In addition to being connected with PRACTICE Deliverable D5.12, this deliverable is also connected with D5.6 "Protocols for the Justification of Risk from Residual Contamination" - which deals with some of the stakeholder issues associated with remediation and related issues – and D7.3 - which provides a procurement protocol for the establishment of Framework Agreements to acquire the necessary equipment and training for preparing for a CB incident.

3.   Timeline

Users of this Tool should note that wastes may arise at various stages following an incident – not just at the remediation stage itself. The potential for waste generation following an incident is shown in Table 1. It is also important to note that negotiations regarding disposal of wastes may be difficult and protracted, especially where the waste volumes are high and/or they contain contaminants which are unusual or for which readily available disposal sites may not have existing authorisations. Waste management issues must therefore be considered from the outset of any remediation planning (see PRACTICE Deliverable D5.12, Annex I: Remediation Plan Template).

Table 1: Incident Timeline, Waste Production and Waste Management Phases

Incident Timeline	Waste Management Phases			
	Waste production	Identification of a suitable site (this document)	Consent for Temporary storage	Temporary Storage and Final Disposal
Emergency response	Some waste will be produced by emergency responders – these may be left packaged at the scene.			
Initial Actions	Initial actions like "making safe" or forensic investigations may produce small quantities of wastes that may be left at the scene.	Information regarding the potential types and quantities of wastes will start to become available and the process of waste site identification can commence		
Remediation Planning Phase (See PRACTICE D5.12)	Some sampling exercises could produce small quantities of waste.		Official consent for the proposed storage arrangements will need to be sought at this stage. Remediation should not start until an appropriate site has been identified.	
Remediation Justification (see PRACTICE D5.12)				
Remediation Execution (see PRACTICE D5.12)	Bulk of the wastes produced in this phase			Wastes transferred to storage as they are produced.
Remediation Confirmation (see PRACTICE D5.12)	Some sampling exercises could produce small quantities of waste.			Some additional waste may be produced from samples. Wastes are stored.
New Normality				Wastes are stored pending final disposition.
Later				Wastes are finally disposed of.

4. Example Scenarios

The following three example scenarios demonstrate some of the issues that may arise for the storage of wastes following CBR type incidents.

1. Chemical Incident at a Conference Centre

    a. **Where:** A Conference Centre – see for example PRACTICE Deliverable D5.12 Remediation Plan Example.

    b. **Cause:** Contamination with a hazardous chemical or similar

    c. **How:** Deliberate release

    d. **Issues:** Potentially large quantities of bulky materials may be produced, which may have been partly decontaminated with oxidising agents to reduce contamination levels. Both the treatment agent used and the remaining contaminant will require specialised destruction; suitable facilities may not be available at short timescales. The wastes will need to be packaged and safely stored pending disposal.

    e. **Example solutions**: Double wrapping of bulk items (seats, carpets etc) in flame retardant polythene and storage in ISO freight containers at secured area of the delivery yard. Storage of polythene wrapped waste in steel drums with polythene liners. Storage of liquid wastes in Intermediate Bulk Containers (IBCs) in locked containers in the yard. Storage of low hazard bulk wastes in sealed skips.

    f. **Within the scope of this deliverable?**: Yes

2. Home made laboratory experiment gone wrong

    a. **Where:** At a residence

    b. **Cause:** Accidental due to incorrect storage of mixtures of chemicals and radioactive materials.

    c. **How:** Careless handling and storage

    d. **Issues:** Storage required pending collection by a suitable specialist contractor for thermal destruction. Mixed wastes, requiring specialist disposal.

    e. **Example solution:** Storage in plastic containers within locked buildings at site.

    f. **Within the scope of** this **deliverable?**: Yes

3. Radiological event in a shopping centre

    a. **Where:** A large multi-tenanted shopping centre in a town

    b. **Cause:** Dispersal of a radioactive powder

    c. **How:** Deliberate release

    d. **Issues:** Potentially extremely large quantities of radioactively contaminated items[8] (e.g. clothing) that will require specialist disposal. Suitable disposal sites may not be available at short notification or they may not be able to accept the waste at the rate of production. On-site storage likely to be contentious. Multiple waste "owners".

e. **Example Solution:** Off-site storage at a disused aircraft hangar which is within a secured airfield. The hangar has a decontaminatable floor finish (originally designed to minimise impact of aviation fuel spills and to withstand aircraft traffic). Drains are fitted with interceptors. Wastes are stored in polythene lined drums or double wrapped in flame retardant polythene.

f. **Within the scope of this deliverable?**: The general approach is applicable but the public engagement/stakeholder issues may require more detailed assessment and some forms of radioactive waste would require very detailed assessment and consideration.

5. The Criteria

The key Criteria are derived from the basic needs that the storage needs to fulfil, namely that the storage is

- Safe,

- Practicable,

- Acceptable and

- Cost Effective.

It is suggested (see MoESDR) that the assessment of suitability of temporary storage sites should proceed in the sequence of "Needs Assessment" to "Logistics Assessment" to "Cost Assessment". These can be seen to map to Safe, Practicable and Cost Effective (respectively) in the model here, but with an extra criterion for Acceptability.

The assessment sequence that is recommended is as shown in Figure 1. The sequence will be cyclic and may require a number of iterations around the criteria until an acceptable overall solution is found.

---

[8] or items that will be assumed to be contaminated because of the difficulty of proving otherwise

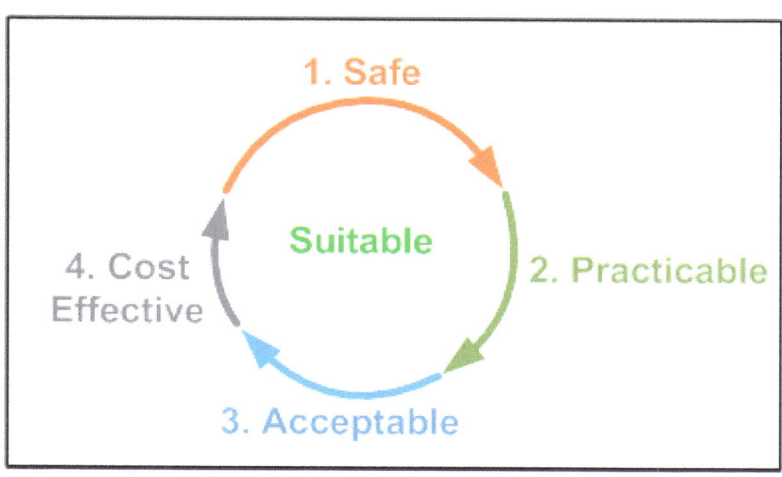

Figure 1: Criteria for Suitability (the length of the arrows provides an indication of the possible relative importance of each the criteria)

These key criteria have been broken down in sub-criteria as shown in Table 2 that have been selected from a review of relevant legislation and literature (e.g. IAEA, MoESDR, SLR Consulting Ltd, ON and COM among others of those listed in Section 7) and also, for the criterion of safety, from a consideration of the types of hazards that may pertain to the types of materials of concern (see Table 3 – developed from BNFL Engineering Ltd and Kletz T - and Figure 2).

Table 2: Sub Criteria

Number	Description
1	**Safe**
1.1	External hazard safe
1.1.1	Safe from extreme external natural hazards
1.1.2	Weather safe
1.1.3	Safe from external man-made hazards
1.2	Passively safe
1.3	Safe from Fire
1.4	Stable
1.4.1	Chemically stable
1.4.2	Physically stable
1.4.3	Thermally stable
1.4.4	Corrosion / erosion stable
1.5	Contained
1.5.1	Contained leakage to air
1.5.2	Contained leakage to ground
1.6	Secure
1.6.1	Secure from people
1.6.2	Secure from fauna
1.6.3	Secure from flora and microbes / fungi
1.7	Isolated
1.7.1	Away from People
1.7.2	Away from other activities
1.7.3	Away from sensitive areas
2	**Practicable**
2.1	Accessible
2.1.1	Accessible for Inspection
2.1.2	Accessible for filling
2.1.3	Accessible for retrieving
2.1.4	Accessible for emergencies

2.2	Observable
2.3	Re-usable
3	**Acceptable**
3.1	To the official group
3.2	To the public group
4	**Cost Effective**
4.1	Reasonable

Table 3: Typical Storage Hazards

Typical Example Storage Hazards		
• Loss of Containment[9]	• Ventilation / Vapours	• Loss of Control
• Fire / Explosion	• Maintainability	• Remote Handling
• Loss of Services	• Effluent or Washings	• Corrosion or Erosion
• Domino[10]	• Extremes of Weather	• Seismic
• Toxicity	• Impact / Drop	• Access or Egress
• Emergencies	• Noise	• Ageing
• Vibration / Movement	• Terrorism	• Ergonomics
• Contamination	• Incompatibility	• Visual

---

[9] E.g., breach of drum, bag, building fabric etc
[10] Failures leading to other failures

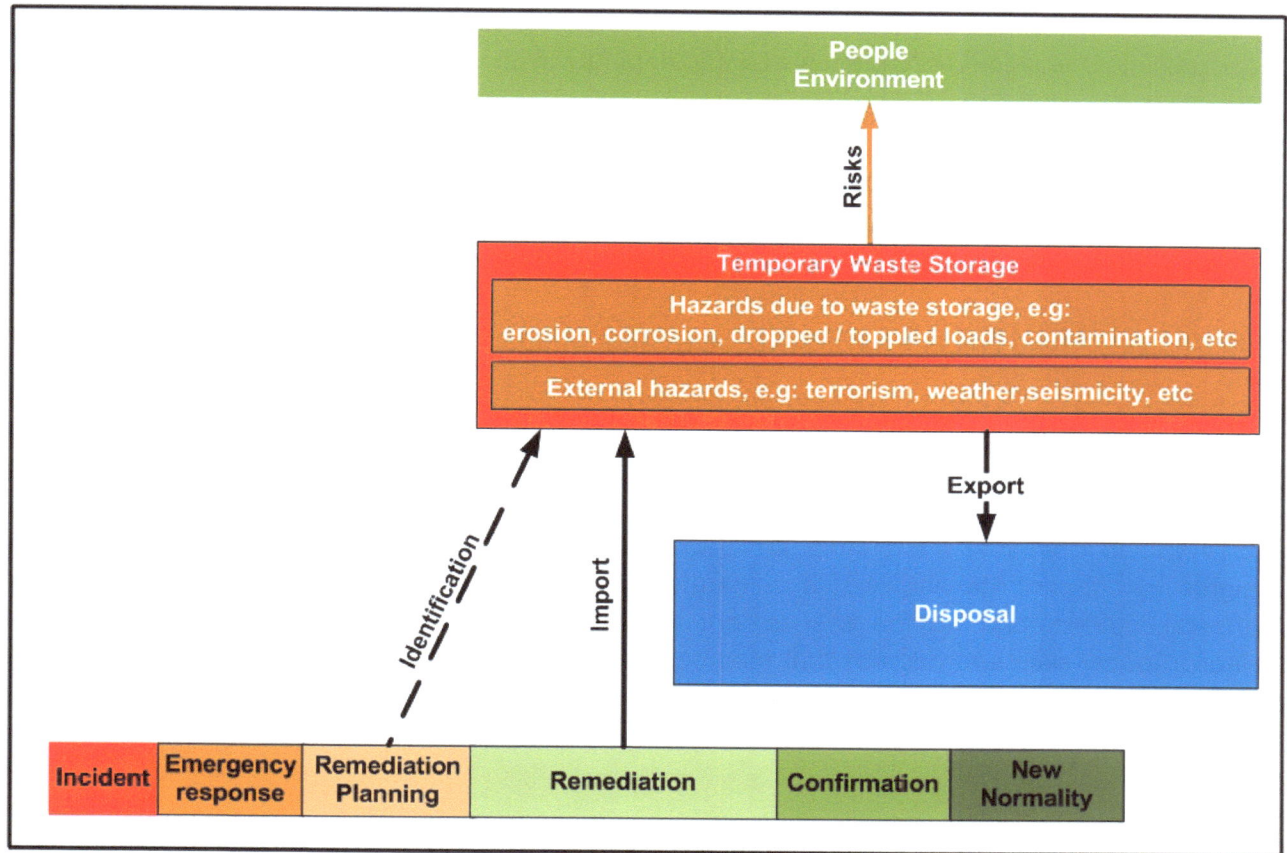

Figure 2: Hazards from Temporary Waste Storage

The resulting criteria and explanatory notes are presented in Annex I

5.1 Compliance with Criteria

The complexity and difficulty of showing that the criteria are met will vary with the types and volumes of wastes concerned; some will be very easy to demonstrate with minimal arrangements (e.g. stable solid wastes might just need to be placed in bags and kept in a locked ISO freight container in a secure compound) whilst others may require more detailed assessment and supporting evidence (e.g. liquid wastes stored in tanks, or wastes that are likely to generate heat or gaseous emissions etc).

Similarly, some of the criteria are factual and are therefore reasonably straightforward to evaluate whilst others are more subjective and may require input from stakeholders. The identification and inclusion of stakeholders is addressed in PRACTICE Deliverable D5.6. (Protocol for the justification of Risk from residual Contamination – see Hale *et al*).

Using the three stakeholder group model used in D5.6, "Protocols for the Justification of Risk from Residual Contamination" the potential stakeholders for waste storage could be as shown in Table 4.

Table 4: Example Stakeholders

Stakeholder Group	Example members
Official Group:	Government, local enforcement agencies, emergency services, security services, investors
Public Group:	Local residents, pressure groups, political groups
Organisation:	Unions, worker representatives,

Users should note that the criteria are each likely to be perceived as having different relative weights by different stakeholders; it is therefore recommended that these weights are discussed with stakeholders prior to the assessment (see PRACTICE Deliverable D5.6 for guidance). A selection of assessment methods that may be used to take account of the different weights can be found in Lehoux *et al* & DCLG.

6.   Conclusions

The management of wastes that arise from remediation is a key issue that needs to be addressed in any remediation project. One of the topics that may require special consideration, early in a project, is the temporary storage of wastes. Such storage can provide a buffer that enables a project to proceed while final disposal arrangements are made.

Criteria have been presented which can be used by organisations wishing to assess the suitability of temporary waste storage arrangements following an incident which will assist in gaining the approval of relevant authorities. The criteria have been developed from a consideration of the types of hazards that may be presented by stored wastes and by examination of existing industry best practice.

The key criteria are that the storage arrangements are;

- Safe,
- Practicable,
- Acceptable and
- Cost Effective.

Sub-criteria have been developed to assist with the assessment.

The importance of stakeholder input into the assessment process has been identified – .especially in respect of judging overall acceptability. The issues are similar to those dealt with in PRACTICE Deliverable D5.6.

7.   Literature

A.T. Kearney Inc for US EPA Region 6 (1997), *Best Management Practices Handbook for Hazardous Waste Containers,* available at www.epa.gov/region6/6en/h/handbk4.pd

Anderson B.L, Sheaffer M.K, Fischer L.E (2000), Hydrogen Generation in TRU Waste Transportation Packages, U.S. Department of Energy, Lawrence Livermore Laboratory. 7000 East Avenue, Livermore, CA 94550

Berney M (2008), *Site Assessment for Waste Management Facilities Hertfordshire County Council,* WSP Environment & Energy (WSPE&E), WSP House, 70 Chancery Lane London WC2A 1AF, UK

(COM) Commission of the European Communities (1999), *Report from the Commission to the Council and the European Parliament on the implementation of Community Waste Legislation,* Com(1999) 752 Final.

(COM) Commission of the European Communities (2003), *Proposal for a Council Decision establishing criteria and procedures for the acceptance of waste at landfills pursuant to Article 16 and Annex II of Directive 1999/31/EC on the landfill of waste,* COM/2002/0512 Final

Corson L.A, Fisher S.A (2009), *Manual of Best Management Practices For Port Operations And Model Environmental Management System,* Great lakes Maritime Research Institute.

(DCLG) Department for Communities and Local Government (2009), *Multi-criteria analysis: a manual,* Department for Communities and Local Government, Eland House, Bressenden Place, London, SW1E 5DU

(EA) Environment Agency (2010), *Non- Waste Framework Directive (NWFD) exemptions - Temporary storage at the place of production,* available at http://www.environment-agency.gov.uk/static/documents/Business/NWFD_3.pdf

Environment Agency, Scottish Environment Protection Agency and Northern Ireland Environment Agency (2011), *Pollution Prevention Guidelines - Drums and intermediate bulk containers: PPG 26* , available at http://cdn.environment-agency.gov.uk/pmho0511btpg-e-e.pdf

European Commission (2006), *Integrated Pollution Prevention and Control Reference Document on Best Available Techniques for the Waste Treatments Industries,* European Commission.

European Commission (various), *Best Available Techniques Reference Documents (BREF),* Joint research Centre Institute for Prospective Technological Studies (IPTS), available at http://eippcb.jrc.es/reference/ .

BNFL Engineering Ltd (2000), *European Commission – nuclear safety and the environment; report on the Magurele Radioactive Waste Treatment Plant, EUR 19259 EN,* available at http://ec.europa.eu/energy/nuclear/studies/doc/other/eur19259.pdf.

(EP) European Parliament (2008),*Directive 2008/98/EC of the European Parliament and of the Council of 19 November 2008 on waste and repealing certain Directives,* Official Journal of the European Union.

Hale N, Kelly D (2012), *Protocols for the Justification of Risk from Residual Contamination,* Project PRACTICE, available at http://practice.fp7security.eu/downloads.

13

HSE (2001), *Reducing risks, protecting people - HSE's decision-making process*, Her Majesty's Stationery Office, St Clements House, 2-16 Colegate, Norwich NR3 1BQ.

(IAEA) International Atomic Energy Authority (2006), *Storage of Radioactive Waste, IAEA Safety Standards Series No. WS-G-6.1*, International Atomic Energy Agency, Wagramer Strasse 5, P.O. Box 100, 1400 Vienna, Austria

Kletz T (1999), *HAZOP and HAZAN; identifying and assessing process industry hazards*, The Institution of Chemical Engineers, 4th ed.

Lehoux N, Vallée P (2004), *Analyse Multicritére*, available at http://www.performance-publique.budget.gouv.fr

(MoESDR) Ministry of Environment and Sustainable Development Romania (2008), *Technical Assistance for the Preparation for Compliance with Provisions regarding Temporary Storage of Waste, Action Plan for Waste Temporary Storage*, available at http://www.mmediu.ro/vechi/departament_mediu/gestiune_deseuri/Action_Plan_for_waste_temporary_storage.pdf

(ON) Österreichisches Normungsinstitut (2008), *Anforderungen an die Ausstattung und den Betrieb von Zwischenlagern für gefährliche Abfälle bei Abfallsammlern nach § 25 AWG 2002*, Österreichischen Wasser- und Abfallwirtschaftsverbandes, Wien

SLR Consulting Ltd (2010), RP 549: *Planning the Processing of Waste arising from a Marine Oil Spill: Part 2: Pre-Incident Planning, Report Reference: 403-02652-00001P2 Version 1*, SLR Consulting Ltd, Aspect House, Aspect Business Park, Nottingham, NG6 8WR

UNEP/OCHA (2011), *Disaster Waste Management Guidelines*, Joint UNEP/OCHA Environment Unit, Palais des Nations, CH-1211 Geneva 10, Switzerland

UNEP, *Mercury INC-4 Draft EU conference room paper*, available at http://www.zeromercury.org/phocadownload/Developments_at_UNEP_level/INC4/Draft_EU_CRP_waste__storage.pdf

(WFD) European Parliament (2008), *Directive 2008/98/EC of The European Parliament and of The Council of 19 November 2008 on Waste and repealing certain Directives*, Official Journal of the European Union.

Criterion	Guidance
1. Safe	The overriding requirement for the storage of any wastes is that it should not present any significant risk to people or the environment. Arrangements that can not be shown to meet this criterion will not be acceptable regardless of how well they meet the other criteria. In general, it will be necessary to consider the likelihood of something adverse occurring and the consequences of it occurring (in terms of impact upon people and the environment) and to show that the resulting risk is acceptable (see Figure 2). **Figure 2: Risk Acceptability** The choice of where the boundaries between Acceptable, Tolerable and Unacceptable are drawn (and their shape) is a matter for each organisation, the relevant authorities and stakeholders. Where the risk is judged to fall in the region that has been deemed to be tolerable it will be necessary to demonstrate that all that can reasonably be done has been done[11]. Risks in the unacceptable region will not be

---

[11] Thus 'tolerable' does not mean 'acceptable'. It refers instead to a willingness as a whole to live with

	acceptable under any circumstances. At the simplest level, Consequence and Likelihood may be subjectively assessed against simple criteria like "Unlikely", "Possible" and "Unlikely" for Likelihood and "Minor", "Moderate" and "Significant" for Consequence. Consideration of the sub-criteria will enable the user to make an informed judgement. For more complex and contentious arrangements a more formal numerical assessment may be required with expert input.
1.1 External Hazard Safe	The storage arrangements must be safe from the effects of external hazards – i.e. those that are generated outside of the storage area by mechanisms unrelated to the storage process itself. In all but the simplest cases wastes should be stored within an overbuilding of some sort.
1.1.1 Safe from extreme external natural hazards (e.g. flood, wind, rain, temperature, earthquake)	The location of the store (and the packaging of the stored wastes themselves) should be protected from reasonably foreseeable external hazards such as extremes of rainfall, wind, temperature and combinations of these. The magnitudes of the extreme hazards that are used for the assessment should be consistent with normal building guidance as a minimum; if the nature and quantities of materials present represents an extreme hazard then greater magnitude events should be assumed. A certain degree of failure of containment may be allowable but the wastes must still be protected from gross loss of containment and the consequences must be shown to be commensurate with the likelihood (see Figure 2). Consideration should be given to the following hazards including their secondary effects noted in {brackets} after each bullet point: • Earthquake / Volcanic activity {loss of power, fire, flood} • River & Coastal Floods {loss of power, landslide} • Extremes of Temperature, rain, snowfall {freezing, collapse, loss of services} • High winds {adjacent building collapse, missiles} • Lightning {fire, loss of services}

The risk in the confidence that the risk is one that is worth taking and that it is being properly controlled. However, it does not imply that the risk will be acceptable to everyone, i.e. that everyone would agree without reservation to take the risk or have it imposed on them (after HSE).

1.1.2 Weather safe	The location should be protected from day-to-day weather. In addition to addressing the issue of extreme weather it is necessary to show that the waste will be protected from normal day to day weather variations. It will also be necessary to show that the waste form will be stable in the envisaged normal daily and seasonal temperature variations.
1.1.3 Safe from external man-made hazards	The storage arrangements must have adequate protection against external man-made hazards – these will usually include items such as; • Missiles (from energetic / pressurised machinery and terrorist attack • Vibration (machinery, vehicles) • Vehicle Impact (road, rail) In the event that very large quantities of waste are to be stored or the potential consequences of loss of containment are great then consideration should also be given to the following hazards: • Aircraft Impact • Explosion • Adjacent hazardous facilities (factories, refineries etc)
1.2. Passively Safe	The storage location and storage method should be passively safe wherever reasonably practicable – i.e. given the intended storage period there should be no need for maintenance, supervision or active protection measures[12] to ensure the continued safety of the materials. The need for monitoring and maintenance to ensure safety should be minimized[13]. There should be no need for prompt corrective action in the event of an incident. Example methods of achieving passive safety are immobilisation of wastes (wrapping in polythene, mixing with concrete / resin), double containment (e.g. double wrapping or wrapping and placing in sealed drums). Passive safety can be demonstrated by first identifying the measures that act to prevent an escalation in consequences and by then considering whether or not any action or intervention is required to ensure continued safety. A passively safe system does not require any such intervention.

[12] An active protection measure is something which responds to a signal or a measurement – which indicates that a hazard is developing - by applying a response which terminates that hazard – e.g. starting a ventilation system in response to rising temperatures.
[13] This does not imply that confirmatory monitoring/inspection is not appropriate, but rather that it is not essential for safety.

1.3 Safe from Fire	The potential for fire should be minimised by the selection of storage materials and method of storage – e.g. the use of non combustible wrappings and drums.

The location should be provided with appropriate fire protection measures; fire fighting equipment materials should be consistent with the chemical nature of the wastes stored (noting that some of the remediation processes used for dealing with CB wastes may themselves be strongly acid or alkali).

Consideration should be given to the potential for waste to generate internal heat and the possibility of spontaneous combustion. This may be especially important where organic materials are present with the potential for bio-degradation. In these cases storage arrangements should allow for sufficient cooling.

Strongly radioactive materials may produce hydrogen through radiolysis of water, moisture and some organic materials present in the wastes. Although this is unlikely to be an issue other than in extreme cases (see Anderson et al) a separate study should be undertaken to ensure that hydrogen levels will remain below safe limits.

A separate fire safety assessment should be produced in consultation with the fire services. |
| 1.4 Stable | The requirement for stability is that the wastes should be in a quiescent state i.e. that they will not significantly change in form or characteristics during the proposed storage period. |
| 1.4.1 Chemically Stable | The location should have an inert atmosphere that will not interact with the waste materials or their storage containers. |
| 1.4.2 Physically Stable | Wastes should be stored in such a way that they will not topple, fall or subside during storage, loading and unloading. Any storage frames and the ground that they rest on should be capable of taking the imposed loads. Drummed wastes should not be stacked as this increases loads on drums and the chances of toppling; storage on purpose designed robust storage racks may be acceptable if it can be shown that dropped and toppled loads present acceptable risks. Drums should be stored upright.

Collection and delivery of loads which may require HGVs or other specialist vehicles (e.g. skips and transporter trucks) should also be considered. |
| 1.4.3 Thermally Stable | The potential for wastes to generate heat should be identified and where it exists it will be necessary to demonstrate that that heat will not lead to rising temperatures and loss of containment. In |

	keeping with the requirement for passive safety (c.f.) heat should be removed by natural means (e.g. convection) without the need for engineered or powered cooling systems.
1.4.4 Corrosion / Erosion Stable	Consideration should be given to interactions between the waste, the containers[14] and their environment (e.g. corrosion processes due to chemical or galvanic reactions). For certain types of waste (e.g. corrosive liquid waste) special precautions should be taken, such as the use of double walled containers, bunding and impervious liners. Where containers are stored in uncontrolled atmospheres (e.g. outside) then they should be suitable for the designated storage period (e.g. have a predicted lifetime that is much greater than the planned storage period) or be made from non corroding material.
1.5 Contained	Wastes must be contained such that i) it does not escape from the storage area by any reasonably identifiable route and ii) it does not cause any uncontrolled secondary emissions or releases.
1.5.1 Contained leakage to air	Where stored wastes are in a form which may lead to airborne hazards in the event of failure of the primary storage arrangements (e.g. bag, drum or package) then the storage arrangements must provide a secondary barrier to the release of these contaminants. In most cases this will be satisfied by double packaging and over-buildings. In extreme cases filtered ventilation systems may be necessary but wastes should be packaged and stored in such a way that they do not produce airborne hazards – unless the packaging fails (e.g. wastes should be stable and quiescent).
1.5.2 Contained leakage to ground	In the event that the primary storage arrangements (e.g. bag, drum or package) fail as a result of any reasonably foreseeable incident, then the storage arrangements should protect against any leakage to ground. For liquid storage this means that some form of secondary containment is provided – with sufficient capacity to encapsulate the loss of containment (typically 100% of the volume). For solid wastes, protection against water ingress and protection of drains (e.g. local drain guards or bunding, or interceptor tanks) should be sufficient. Adjacent surfaces should be water proof or protected with impermeable membranes.
1.6 Secure	Secure waste storage means all reasonable precautions are taken to ensure that the waste cannot escape from the storage and members of the public are unable to gain access to the waste.

---

[14] The choice of "container" form and material of construction is principally determined by the waste characteristics but it is also affected by the envisaged storage period and the proposed storage arrangements. Some guidance eon packaging of wastes can be found in AT Kearney Inc.

	(adapted from EA)
1.6.1 Secure from People	Security against unintentional or deliberate intrusion should be provided – especially where the materials may contain hazardous materials that may still be of interest to terrorists or activists. As appropriate areas should be lockable and capable of being monitored and observed. Security and access controls may be required to prevent the unauthorized access of individuals and the unauthorized removal of waste material. The level of security and access control required at a waste storage facility should be commensurate with the hazards from and the nature of the waste. Measures may range from simple locked storage containers/areas to CCTV monitored/security patrolled buildings.
1.6.2 Secure from Fauna	Wastes should be stored in containers which keep out rodents and insects – which may otherwise lead to a spread of contamination and a loss of containment. If this is impracticable then the storage location itself should be secure against fauna. This may be especially important if the wastes contain any organic or putrescent materials.
1.6.3 Secure from Flora and microbes/ fungi	For storage periods which may extent to several months, the storage arrangements should be secure from ingress by plants, microbes and fungi – which may breach the containment and cause contaminants to leach out.
1.7 Isolated	The criterion of isolation is to ensure that if anything does go wrong with the storage arrangements then the potential for that to lead to exposure of people and the environment is minimised by them being suitably distant from the storage.
1.7.1 Away from People	Storage sites should be as far as is reasonably practicable from people – other than those involved their operation - and every day activities. The degree of isolation required will depend upon the magnitude of any hazards arising from potential failures in storage arrangements or waste handling[15].
1.7.2 Away from other activities	Storage areas should be isolated from other activities and movements as far as reasonably practicable. This criterion ensures that there is minimal potential for confusion regarding the waste storage arrangements and reduces the potential for accidental mixing of wastes or incorrect sentencing.
1.7.3 Away from	Storage areas should be chosen such that they are not near to

---

[15] Since the intensity of the radiation falls with distance, the requirement for isolation may be especially important in the case of radioactively contaminated items.

Sensitive Areas	sensitive receptors and they should be in environmentally stable areas, i.e. they should not be in environmentally sensitive areas such as floodplains, wetlands or groundwater sensitive areas. This criterion limits the potential environmental impact of any failures in storage arrangements or waste handling.
2. Practicable	**Once the criterion of Safety has been considered and satisfied, the practicality of a site can be considered in terms of whether it is workable, given any constraints imposed by the safety criteria - i.e. "does the site have the other physical characteristics that are needed to ensure that it can be operated?"**
2.1 Accessible	Waste must be accessible at all stages of storage and at all times.
2.1.1 Accessible for inspections	Wastes should be stored such that they remain accessible for routine inspection or in the event that any packages exhibit any signs of failure. This criterion becomes increasingly significant, as waste volume and waste storage duration increase. Access to some wastes for inspection may lead to the requirement for change rooms and monitoring stations. Although these may be mobile – rather than permanent features – the storage arrangements will need to provide sufficient space for such facilities.
2.1.2 Accessible for filling	The storage arrangements must compatible with the proposed filling arrangements – i.e. the order in which packages may arrive, their physical size and shape. Consideration should also be given to the potential number and type of vehicle movements that will be required.
2.1.3 Accessible for retrieving	Clearly the arrangements must be compatible with the need to eventually empty the temporary store and dispose of the wastes. There should be nothing in the storage arrangements that will restrict eventual emptying of the store. Different types of wastes should be segregated and separate from each other (e.g. radioactive from bio-contaminated wastes); "segregation" implies that physical barriers[16] exist between the different wastes types (e.g. walls, over-packs, gratings etc) and "separation" implies that the packages are separated by a suitable distance. Segregation and separation of wastes will help to avoid cross-contamination and accidental removal from the controlled

---

[16] In the case of radioactive wastes, the physical barriers may be required in order to provide shielding to protect people.

	environment. It may also reduce the exposure of workers in normal operations and may limit the severity of any consequences under accident conditions (adapted from IAEA).
2.1.4 Accessible for emergencies	The arrangements must permit the prompt removal of any wastes packages that show signs of failure, without undue movement of other packages. Similarly, it should be possible to clean-up spills and local loss of containment incidents without undue disturbance of other intact wastes[17].
2.2 Observable	The storage arrangements should permit on-going monitoring and routine inspection of the stored wastes. Storage arrangements must provide adequate access for visual inspection and monitoring without the need to move wastes.
2.3 Re-usable	Storage areas and arrangements should ensure that once emptied the stores can be returned to unrestricted use / service. Arrangements must therefore ensure that in the event of an incident leading to the spread of contamination, the storage areas can be decontaminated – e.g., polythene linings and finishes that can be readily decontaminated can be used.
3 Acceptable	**Meeting the criteria listed for Safety and Practicality will ensure that most stakeholder concerns will be addressed. However, stakeholders must be explicitly included in these decisions.**
3.1 To the official group (see 5.6)	Given the nature of some of the waste materials that may be stored, it will be appropriate to consult with local agencies (e.g. local licensing authorities, police, ambulance, fire services) to ensure that their requirements (either explicit or implied) will be met by the proposed storage arrangements. Their views should be canvassed before any application for storage permits is made.
3.2 To the public group (see 5.6)	The public group's potential concerns will be addressed to a large degree by conforming to the criteria presented earlier – especially those regarding isolation - but it may be appropriate to include a representative public stakeholder group in the selection process. PRACTICE Deliverable D5.6 "Protocols for the Justification of Risk from Residual Contamination" (see Hale *et al*) provides guidance in respect of such processes.
4 Cost Effective	**Cost effectiveness will largely be a commercial matter for the organisation wishing to store the wastes to consider. Cost effectiveness should only be used as a screening criterion for options which satisfactorily meet the other criteria. Both**

---

[17] Spill kits should be provided at the storage locations.

	Operating and Transport costs need to be addressed.
4.1 Reasonable	Organisations will need to show that they have provided reasonable amounts of funding to meet Criteria 1 to 3. The definition of 'reasonable' in this instance will be a matter for not just the organisation but also key stakeholders such as insurers and the Authorities.

www.ingramcontent.com/pod-product-compliance
Lightning Source LLC
Chambersburg PA
CBHW050434180526
45159CB00006B/2538